BUILDING BLOCKS OF BIOLOGY

ANIMALS

Written by Alex Woolf

Illustrated by Ruth Bennett

WORLD BOOK

www.worldbook.com

Co-published by agreement between Shi Tu Hui and World Book, Inc.

Shi Tu Hui
Room 1807, Block 1,
#3 West Dawang Road
Chaoyang District, Beijing 100025
P.R. China

World Book, Inc.
180 North LaSalle Street
Suite 900
Chicago, Illinois 60601
USA

WORLD BOOK STAFF

Editorial

Vice President
Tom Evans

Senior Manager, New Content
Jeff De La Rosa

Proofreader
Nathalie Strassheim

Graphics and Design

Senior Visual Communications Designer
Melanie Bender

Library of Congress Control Number: 2025942737

Building Blocks of Biology
ISBN: 978-0-7166-6737-7 (set, hard cover)

Animals
ISBN: 978-0-7166-6741-4 (hard cover)

Also available as:
ISBN: 978-0-7166-6761-2 (e-book)
ISBN: 978-0-7166-6751-3 (soft cover)

Acknowledgments
Writer: Alex Woolf
Illustrator: Ruth Bennett/The Bright Agency

TABLE OF CONTENTS

There is a glossary on page 40. Terms defined in the glossary are in type **that looks like this** on their first appearance.

WHAT IS AN ANIMAL?
One pleasant afternoon...
Fin! Fur! What a perfect day for a walk in the countryside!
You said it, Root! These woods are full of life. See that squirrel!
And look at those flowers!
There are so many animals around, Fin!
Yeah, but... I have a silly question...

There's no such thing! Ask away.
What exactly is an animal?
I mean, what makes it different from a plant, for example?
They're both living things, aren't they?

You raise a great question, Fin! There are actually many different kingdoms of living things...
Plants
Animals
Fungi
Animals, plants, and fungi are made up of many cells.
Protists
Bacteria
Archaea
Protists, bacteria, and archaea are made up of just a single cell.

What do you think makes animals different from plants and fungi?
Well, animals can move around, for one thing...
Furball is right! Most plants and fungi have roots or rootlike structures that keep them... well... "rooted" in place.
Animals can move around with help from their legs or wings...
Or fins!
Or, they just use their bodies' muscles to inch along.

Why do you think animals need to move?
So they can find food!
Hey! Be careful!
Sunlight
CO_2
Water
That's right! We plants don't need to move to find food... We make our own!
Plants make their own food using water from the soil and carbon dioxide from the air. They combine these to make simple sugars using the power of sunlight, in a process called photosynthesis.

Oxygen
Respiration
Photosynthesis
Plants and animals breathe differently.
CO_2
In photosynthesis, plants like me take in carbon dioxide and emit oxygen...
Animals like us breathe in oxygen and breathe out carbon dioxide!
That's why we all need each other!!!
Identify two differences between animals and plants. See page 40 for answers.

CLASSIFYING ANIMALS

It's a way of organizing living things into different groups based on their similarities.
It makes them easier to identify and study.
The main division in the animal kingdom is between...
Vertebrates- animals with a backbone and...
Invertebrates- animals without a backbone

There are many different kinds of vertebrates, each with its own characteristics.
Mammals like me are warm-blooded.
We have fur or hair and produce milk to feed our young.
Fish like me are cold-blooded and are covered in scales.
We have gills to breathe underwater and streamlined bodies.
Birds are warm-blooded. They have feathers, beaks, and wings and lay eggs.
Reptiles are cold-blooded and covered in scales. They mostly lay eggs.
Amphibians are cold-blooded with moist, smooth skin. Most live part of their life in the water and part on land.

There are also many different kinds of invertebrates.
Insects have six legs, three body segments, and antennae.
Arachnids have eight legs and two body segments.
Annelids have long, soft, segmented bodies.
Crustaceans have hard shells, jointed legs, and antennae.
Mollusks are soft-bodied creatures with no legs. Some develop hard shells for protection.

Now, let's see if we can identify your mystery animal, Fin.
How many legs did it have?
Four.
Could it be an invertebrate?
Insects have six legs.
Arachnids have eight legs.
Millipedes have over 1,000 legs.
Annelids and mollusks have no legs.
None have four legs.
The mystery animal is probably not an invertebrate.

If it has four legs, it can't be a fish or a bird...
Did it have fur?
No.
No fur
Hmm! So, it's a vertebrate...
Four Legs
MAMMALS
FISH
BIRDS
AMPHIBIAN
REPTILE
VERTEBRATES
INVERTEBRATES
It has four legs and it doesn't have fur...
So, it must be a reptile or an amphibian.
Is a fly a vertebrate or an invertebrate? How about a snake? A snail? A person? See page 40 for answers.

SCIENCE FUN WITH FUR!

ANIMAL CLASSIFICATION FLOW CHART

YOU WILL NEED:

- A large sheet of paper
- A pencil or pen
- Pictures of animals printed off the internet or cut out of magazines

1. Copy the chart on the facing page onto your sheet of paper.

(2) For each animal, start at the top of the chart. Move down the chart as you answer each question, until you get to the place where your animal belongs.

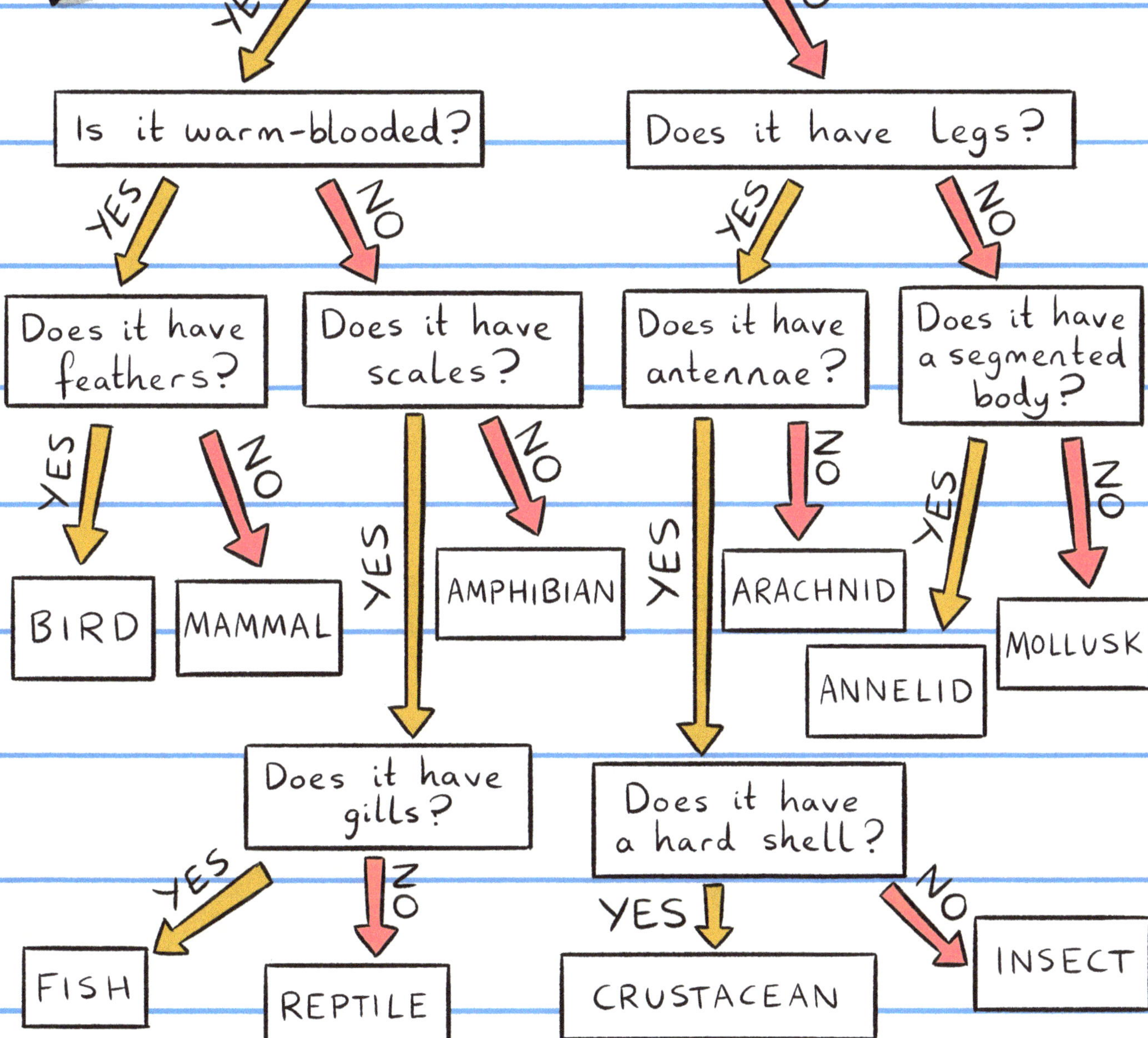

LIFE CYCLES
Did the mystery animal look young or old, Fin?
How should I know? I didn't ask it for ID!
Animals look different at different stages of their lives.
It's true. Animals go through different stages from birth to adulthood—called the **life cycle.**
Take a butterfly, for example. It starts its life cycle as an egg.
A female butterfly lays eggs on the leaves or stems of plants.
The egg hatches into a caterpillar. As it grows, the caterpillar sheds its skin. Underneath, it has a new, bigger skin.

HIC
Look, it's eaten a huge hole in the leaf!
What's happened to the caterpillar?
It's changed into a pupa.
During the pupa stage, the caterpillar hangs upside down in a cocoon or chrysalis.
Look! Something's coming out of the cocoon!
It doesn't look like a caterpillar anymore.
Inside the outer covering of the pupa, the caterpillar transforms into a butterfly.
Now the butterfly is ready to emerge and spread its beautiful wings!

But the mystery animal isn't a butterfly —it's either an amphibian or a reptile!
True! Maybe we should look at one of their life cycles.
A frog is an amphibian. It has an interesting life cycle.
A frog starts its life cycle as an egg.
Adult frogs lay hundreds of eggs that clump together in groups called frogspawn.

The baby frog that hatches from the egg is called a tadpole.
The tadpole lives in water. It has gills and a long tail for swimming.
Next, the tadpole grows legs.
Then, its tail shrinks away, and skin grows over its gills.
The frog develops lungs for breathing air.
The frog is now fully grown.
I don't think the mystery animal was a frog—young or old.
Put these stages of a butterfly's life cycle in order, beginning with the egg: egg, adult, pupa, caterpillar.
See page 40 for answers.

FOOD CHAINS

What was the mystery animal doing, Fin?
?
It was eating a bug...
...and I think it was being hunted by a bird.
We can learn a lot about an animal from its place in the **food chain.**
What's a food chain?
Is it a chain you can eat?

No, it's really all about energy. All living things need energy.
A food chain describes the flow of energy from one living thing to another.
Plants get energy from sunlight.
Some animals get energy from eating plants or plant materials.
Some animals get energy from eating other animals.

Plants are at the base of most food chains.
They are called **producers** because they make their own food.
An animal that eats plants is called a **primary consumer**—this bug, for example.
Primary consumers are the next stage of a food chain.
The mystery animal ate a primary consumer...
So that makes the mystery animal a **secondary consumer.**
?
Secondary consumers (often called *predators*) kill and eat other animals.

The animal at the top of a food chain is called the **top predator.**
Changes in a food chain affect every living thing in it.
For example, if plants become scarce, animals will have less to eat... all the way up the chain!
What does a primary consumer eat? What does a secondary consumer eat? See page 40 for answers.

ECOSYSTEMS
Where exactly was the mystery animal, Fin?
It was over there by the pond.
So, it must have been part of the pond's **ecosystem.**
The eco... eco... what?
The ecosystem! It just means a community of living things and the nonliving things they depend on in a particular area.
Rain forests, coral reefs, and deserts are all examples of ecosystems.

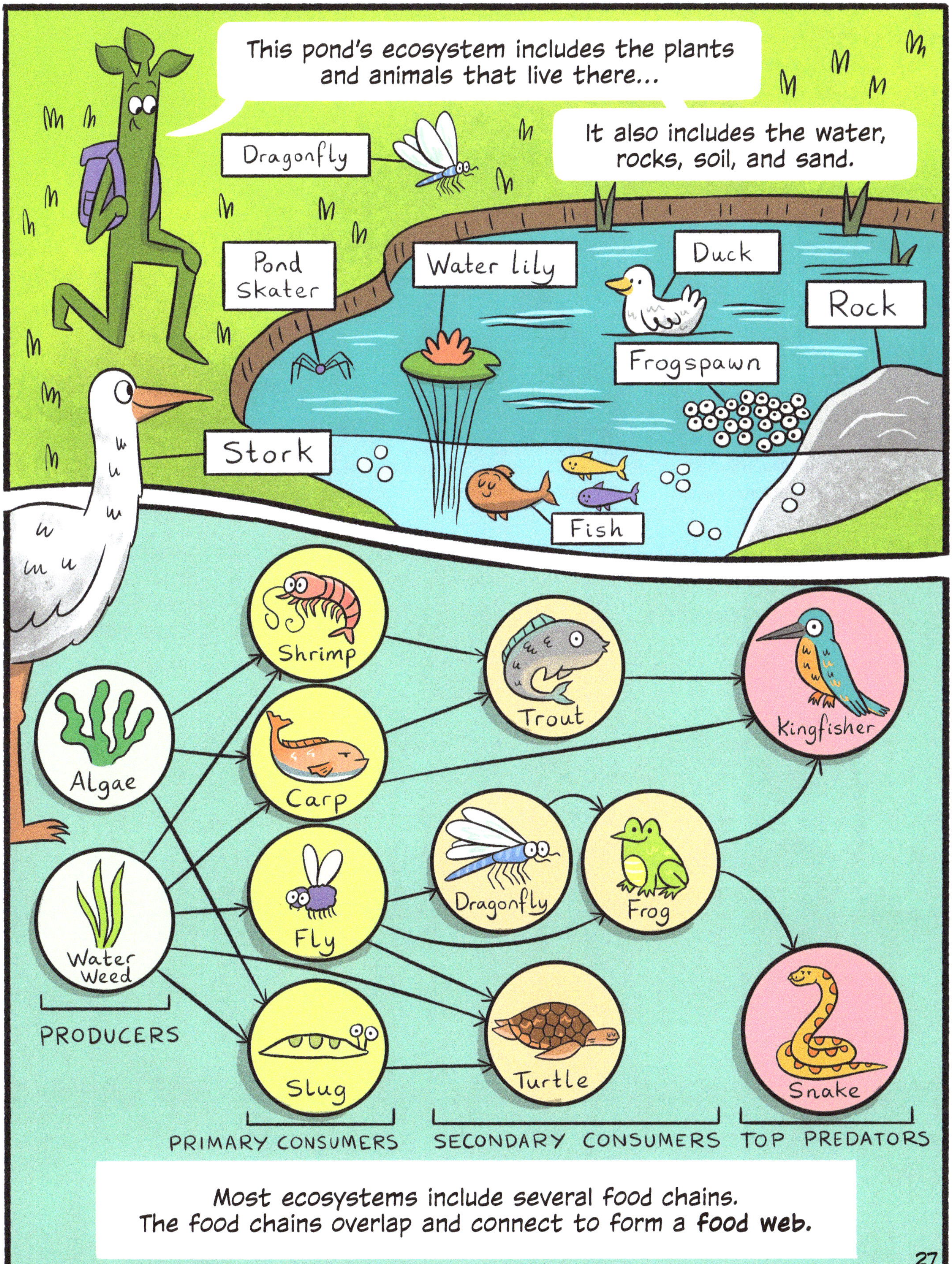
This pond's ecosystem includes the plants and animals that live there...
It also includes the water, rocks, soil, and sand.
Dragonfly
Pond Skater
Water lily
Duck
Rock
Frogspawn
Stork
Fish
Shrimp
Trout
Kingfisher
Algae
Carp
Water Weed
Fly
Dragonfly
Frog
Slug
Turtle
Snake
PRODUCERS
PRIMARY CONSUMERS
SECONDARY CONSUMERS
TOP PREDATORS
Most ecosystems include several food chains.
The food chains overlap and connect to form a **food web.**

Even death plays an important role in maintaining a healthy ecosystem.
When animals die, earthworms, fungi, and bacteria help their bodies to decompose (break down).

These **decomposers** also break down wastes and dead plant matter, returning nutrients to the soil.
Plants (producers) use these nutrients to grow, providing food for animals (consumers).
Animal eats plants
Animal dies
Decomposition
Nutrients added to soil
Plant grows

Producer
All the living things in an ecosystem depend on one another.
Therefore, ecosystems often exist in a delicate balance.
If one kind of living thing dies out or grows too plentiful, everything else is disrupted.
Producers
Decomposers
Consumer

?
It looks like my mystery animal is important to the pond!
It relies on the other living things there...
And they rely on it!
Fill in the blanks. An ecosystem is a community of _____ things and the _____ things that they depend on. See page 40 for answers.

ANIMAL ADAPTATION
There's something else I forgot to say about my mystery animal...
What's that?
It was quite hard to see.
Did it blend into the background like this butterfly?
What butterfly?
Exactly!

Another word for blending in is **camouflage.**
Camouflage is an example of animal **adaptation.**
You mean animals must adapt to their environment if they are going to survive?
Yes! Look at this camel, for example.
hump for storing fat
long eyelashes and thin nostrils that close to protect against blowing sand
leathery mouth for chewing thorny desert plants
wide feet for walking on soft sand

Camouflage is an important adaptation.
It helps prey animals, such as this squirrel, hide from predators.
And, it helps predators like this jaguar sneak up on their prey.
Decorator crabs decorate their shells with seaweed and bits of sponge to blend into their environment.

Did you spot the mystery animal? See page 36 for the answer.

LIFE on the EDGE
Ideas from the Cutting Edge of Biology
WILDLIFE-TRACKING DRONES
Hey, what are you up to, Root?
I'm tracking elephants using this wildlife drone. Do you want to look?
WHIRRRRRRRRR
Why are you doing this?
Elephants are endangered. By tracking their movements, we can help protect them.

Tracking animals can help us learn about their populations, lifestyles, habitats, and migration patterns.
Learning how they interact with their environment helps with conservation efforts.
How does it work?
BEEP
BEEP
The animals have collars that send radio signals that are picked up by the drone.
A drone can track up to 40 animals at the same time.
It can track them through forests and rugged landscapes.
Drones can even use thermal imaging to track untagged animals using their body heat.
Neat!

SOLVING THE MYSTERY
There it is! The mystery animal!
I see it!
So do I! Let me take a photograph.
It has smooth skin, not scales.
It has four legs and a tail... and it lives close to water.
This makes me think it's an amphibian.

Let me use an app on my phone to identify it...
5127
CLICK!
A few minutes later...
It's a newt!
The mystery is solved! Hooray!
NEWT

SHOW WHAT YOU KNOW

1. Are each of the following vertebrates or invertebrates?

A. mammals
B. insects
C. annelids (worms)
D. reptiles
E. amphibians

2. Put this food chain in order, beginning with the producer:

secondary consumer; producer; top predator; primary consumer

3. Is each of the following a characteristic of plants, animals, or both?

A. moves around to find food
B. adapted to living in its environment
C. takes in carbon dioxide and gives off oxygen
D. depends on living and nonliving things in its ecosystem
E. must breathe in oxygen to live

4. True or False?

A. An ecosystem consists of only the living things in a particular place.
B. Decomposers help break down dead plant and animal matter.
C. Animals make their own food.
D. Most amphibians live their lives on land.
E. Camouflage is an adaptation for blending in with an organism's environment.

See page 40 for answers.

ANSWERS

page 9: (1) Animals generally move to find food to eat. Plants make their own food using the energy in sunlight. (2) In photosynthesis, plants take in carbon dioxide and produce oxygen. Animals breathe in oxygen to make energy and give off carbon dioxide.

page 15: fly- invertebrate; snake- vertebrate; snail- invertebrate; person- vertebrate

page 21: egg, caterpillar, pupa, adult

page 25: A primary consumer eats producers. Secondary consumers eat primary consumers.

page 29: living; nonliving

SHOW WHAT YOU KNOW ANSWERS
pages 38-39:

1. A. vertebrates
   B. invertebrates
   C. invertebrates
   D. vertebrates
   E. vertebrates
2. producer ⟶ primary consumer ⟶ secondary consumer ⟶ top predator
3. A. animals
   B. both
   C. plants
   D. both
   E. animals
4. A. False. An ecosystem also includes nonliving things that living things depend on.
   B. True.
   C. False. Animals must eat food made by plants or other producers.
   D. False. Most amphibians live at least part of their lives in the water.
   E. True

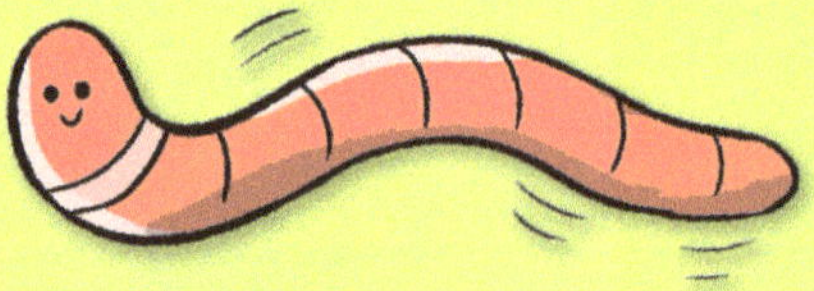

WORDS TO KNOW

adaptation a trait that helps a living thing to survive in its environments.

amphibians cold-blooded animals with moist, smooth skin that often spend part of their lives in the water and part on land.

camouflage an adaptation for blending into a particular background.

decomposer an organism that breaks down wastes and dead organic material, helping to return nutrients to the ecosystem.

ecosystem a community of living things along with all the nonliving things they depend on in a particular area.

food chain the flow of energy from one living thing to another through food.

food web a network of food relationships consisting of multiple overlapping and interconnected food chains.

invertebrates animals without a backbone.

life cycle the changes an organism undergoes as it grows, develops, and reproduces.

mammals warm-blooded animals with hair or fur that feed their young on mother's milk.

primary consumer a living thing that gets food from a producer, for example an animal that eats plants.

producer an organism that makes its own food.

reptiles cold-blooded animals covered in scales that mostly lay eggs.

scientific classification a way of organizing living things into categories, based on their characteristics and shared ancestry.

secondary consumer an animal that eats other animals, also called a predator.

top predator the highest consumer in a food chain or web; an animal that preys on other animals but is not generally preyed upon.

vertebrates animals with a backbone.

www.ingramcontent.com/pod-product-compliance
Lightning Source LLC
LaVergne TN
LVHW070859210926
848482LV00016B/973